THE ABC
OF
INFORMATION SECURITY
PART ONE
Timothy Asiedu.

authorHOUSE®

AuthorHouse™ UK Ltd.
500 Avebury Boulevard
Central Milton Keynes, MK9 2BE
www.authorhouse.co.uk
Phone: 08001974150

First published by AuthorHouse 11/23/2011

ISBN: 978-1-4567-7436-3 (sc)

INTRODUCTION

Information Security plays very useful role in the day to day management of our Computer Systems. This book on Information Security, which is packed with a lot practical issues, tends to make the profession as easy as ABC. In going through the materials explained in the book, you will come across many issues you face in the business environment.

The book is recommended for everyone in the business environment, especially IT Professionals, Chief Executive Officers, Managers and other Computer users. The book is full of examples to make the study of Information Security as easy as possible. Most of the materials treated in this book were issues encountered at the workplaces of the author and also through his research studies. Based on the experiences in Information Technology of the author, the material in the book has been presented in a way to make Information Security very interesting.

Chapter one of the book focuses on basic Information Security definitions which you will normally encounter in your day to day management of the Business environment.

Chapter Two covers issues on Basic Information Security concepts. After defining the Basic Information Security Concepts, thus Confidentiality, Integrity and Availability; the author follows with the Dos and Don't of Information Security.

Chapter Three focuses on the Causes of Security Failures. These areas are quite critical since it can serve as a warning to all of us. Once issues are clearer, we will be in a position to guard against what we may be encountering in our various offices.

In chapter Four, readers will learn how to develop a Corporate Information Security Policy. In fact, Information Security Policy is very important for any organization embarking on Information Security. Issues regarding Standards and procedures in the deployment of Information Security in the Business environment can be found in the Information Security Policy handbook which all employees in an organization are supposed to have a copy. Samples of Information Security Policy and also E-mail Policy have also been included in this book.

Chapter Five concerns with Data Security Audit. Data Security Audit normally cross check whether we have managed to put in place good procedures and standards in our business environments. Understanding the good concepts and procedures are fine, but we have to be very sure whether the implementation has been done properly. Through Audit it can be established whether what you claim has been done has actually been done properly.

Chapter Six dwells on User Access Control. In accessing the Computer System, what available controls and standards have the author written about. You need this part one of the ABC of Security to know what the author has for you.

This part of One ABC of Information Security will soon be followed by the Part Two of the book. In part two, the author will write on securing our Networks. This will also include security issues on the Internet.

DEDICATION

To my Father: Daniel Anka Asiedu, the wisest man I knew

To my Mother: Emily Osam, the most caring woman I know

To my Wife: Esther, the most loving woman I know

To my two daughters: Andrea & Christabel, the smartest girls I have come across.

CONTENTS

CHAPTER ONE
DEFINITIONS:

What is Information Security ?:

Information Security is about the security of our Information System on our PCs in the offices & others. Observing basic rules so that, you can continue to run your computer systems in the office & other places is all about Information Security. Information Security does not end in ensuring protection of our Information System on the PCs but it goes further to cover the protections of documents we use in the offices.

The Role of Information Security Manager:

In today's world of abundance of information, it is always appropriate you appoint a Manager to look after the co-ordination of Information Security in your offices. That manager must have a direct report to the Chief Executive Officer (CEO) of the organization. On a continual basis, this manager will have to send Information Security report to the CEO.

What is Information Security Policy? :

Documented policies of Information Security are normally put together in handout for use by all Computer Users in an organization. The local policy document must contain a statement of management intention supporting the goals and principles of Information Security.

What is Internet Fraud?

Electronic Fraud is the piracy of Information for pecuniary and other benefits. Thus any act carried out with criminal deception using Information technology as a medium constitutes electronic fraud.

Self – Assessment Questionnaire:

Self – Assessment Questionnaire plays useful role during Data Security Audits. It normally contains prepared questions on Information Security used during the audit of our Information Systems and its environments.

Information Owners, custodians:

Every Information System must have an owner and this owner further appoints an Information Custodian who then acts in his/her absence.

Information Classification:

Information can be classified as follows:

    i)   Confidential
    ii)   Restricted
    iii)   Internal Use
    iv)   Public

Confidential information is supposed to be handled only by the owner of that information.

Restricted Information is normally used within a particular group, like the marketing or Information Technology department of an organization.

Information is marked as Internal Use only if that information is supposed to be used by the entire organization but not public.

Lastly, Public Information is normally meant for a wider group, thus those personnel outside and inside your organization. It is not restricted to any group.

Security Incidence:

Reported case of Information Security lapses in an organization constitutes a security Incidence. The following are some Security Incidence:

- Virus Incidence
- Theft of PC's or components
- "Chain Letters" circulating on e-mail system.
- Power failure
- Unauthorized login account established with supervisor rights.
- Unauthorized access attempts by external personnel.
- Water Damage

Security Breach:

A Security breach is said to occur when information Security measures have failed to work as required for a particular situation. Assuming you have been advised by the Information Security Manager not to bring to the office any foreign disk you have been working with at home and you disregard that and bring in that foreign disk to the office and a problem like virus infection occurs; this is normally referred to as a Security breach.

Firewall:

Firewall is normally a security systems put in place to check intruders or hackers from breaking into corporate Network.

Intrusion Detection System (IDS):

Application that continuously monitors the network traffic and the operating Systems for any attacks and breaches is referred to as Intrusion Detection System. Its main goals are to report on any security lapses occurring on a running organization network.

Encryption:

Encryption is a means by which information is prevented by an unauthorized access by rendering these communications illegible to unintended viewers. In fact the approach is to maintain a free and connected structure of networks and Internet by allowing data to flow freely in a secret language known only by the sender and the receiver.

Virus and anti-virus software

Computer virus is destructive in nature and normally comes to your com-puter unknowingly from the Internet or other infected item like the disk. This virus normally hides on the Computer and like a time bomb strike by attacking a data file or other part of the computer. Some even can go to the extent of erasing files from the hard disk. You will normally require anti-virus software like McAfee, Kaspersky or Norton anti-virus software to be used to clean the virus. Example of a virus is the Stone virus.

Similar to the virus is the spyware, which can also come to your computer from the Internet. They normally hide on your computer and with time can disable certain Internet sites that you visit often. When that Internet site is disabled, you will not be able to visit that site until the spyware is cleaned with anti-spyware software from your PC.

Confidential Undertakings:

All employees of an organization need to sign a confidentiality undertaking, agreeing to safeguard company information and information assets whilst in employment or some time afterwards.

Help Desk:

A table set-up in the I.T Department of the organization which Computer Users turn to for assistance or solution when they come across a problem.

Chapter Two

Basic Information Security Concepts & Dos and Don't

Under the Basic Information Security Concepts, the three main qualities of information are handled. They are as follows:

a) Confidentiality
b) Integrity
c) Availability

Confidentiality:

Under confidentiality, we concern ourselves with the assurance that Information is not disclosed to unauthorized persons inside or outside the company and that authorized release of information will not cause any loss or business disadvantage.

Integrity:

What is Integrity under Information Security? Integrity gives the assurance that information accurately represents the authorized business activities of your company and it is not corrupted or modified by unauthorized persons inside or outside the company.

Availability:

Under availability, we have the assurance that information and information Systems will be available when required by the company's business needs or to comply with regulatory or legal requirements for information disclosure.

Information is so valuable that it is always appropriate you observe the above basic principles of Information.

The following are some of the **Do** and **Don't** of Information Security.

DO

i) Do ensure that your Personal Computer is protected by approved and up-to-date anti-virus software. Examples of some of the anti-virus software are McAfee and Norton Anti-Virus software.

ii) Do keep your password secret. In case you allow anyone to use your password you may be held responsible for his/her action.

iii) It is appropriate to change your password regularly – at least once each month. Try as much as possible to select passwords which are difficult for other persons to guess.

iv) Use only properly purchased and licensed computer programs or software.

v) It is always advisable to make sure that all computing equipment is secured against loss or theft.

vi) Use, and respect, the approved security classification markings on documents. The approved classification markings are as follows:
   - Confidential
   - Restricted
   - Internal Use and
   - Public

vii) It is always appropriate to collect output from printers promptly.

viii) You must always ensure that you clear your desk of documentation at the end of day, and lock up desks and cabinets. This is what we call "Clear Desk" Policy.

ix) Report suspicious or unusual system behaviour to your Country Help-desk. In case of any suspected Computer virus, shut down the system immediately and call the Help-desk.

x) You should ensure that the frequency of back-up is adequate for your business needs for information recovery. Normally we keep three sets of the back-ups, Grandfather, Father and Son.

Don't

i)   Do not write down your password or let another person see you using it.

ii)   You shouldn't let another person use your password or your logged-on system.

        Do not ask another system user for his/her password or use another person's logged-on system.

iii)   Do not make unauthorized copies of software.

iv)   You should ensure that, disk or sensitive computer output is not left lying around – put them away safely. Ensure that printers are cleared of unused print-outs.

v)   Do not allow non-employee, e.g. consultants or sub-contractors, to have access to your company's computer equipment or data unless they can show written authority from any of your company's manager.

vi)   Ensure that before you leave your desk in the office, you should log-off your computer. Also ensure that you have approved screen – savers with password protection.

vii)   Do not use disks on your company's PC unless they have been checked of virus and authorized for use on your system. In case you want your disk checked, contact your Help – desk.

viii)   Never leave portable computers, disks or tapes containing sensitive information unattended or out of your control, e.g. in hotel rooms, luggage racks, or vehicle.

        Keep them with you at all times, or lock away in a safe place.

The above list is not completely exhaustive. It can grow with time.

CHAPTER THREE
CAUSES OF SECURITY FAILURES

The following are some causes of Security Failures:

i)   Over-reliance on key staff
ii)   Staff Dissatisfaction
iii)   Inadequate documentation – In development
                                                  - In maintenance
iv)   Unauthorized program changes
iv)   No record of program/master file amendments
v)   Inadequate program backup
vi)   Inadequate user controls
vii)   Bad physical security

i)   Over – reliance on Key Staff :

Do not over-rely on your key staff, otherwise the day he/she leaves your organization the effect will be disastrous. It is always appropriate you develop the young ones in the various departments, so that they can step into the shoes of the key staff whenever he/she   leaves the organization.

You never know when the key staff will be dissatisfied and thus leave your organization. For you to have business continuity, you are supposed to have a backup staff. Not only the key staff leaving the organization but can also fall sick. Always prepare for the unexpected.

ii)   Staff Dissatisfaction:

Staff Dissatisfaction can easily lead to a loss of key staff and thus goes to affect the business. In order to bring down the level of staff dissatisfaction, as a business manager it is always appropriate you study your staff behaviour properly and thus try as much as possible to meet your staff expectation in a way. They are the life blood of your organization, without them you do not have an organization.

As a Business Manager, have you ever sat down to analyze why staff get dissatisfied? Many staff get dissatisfied due to poor condition of Services. The following are some of the poor conditions of Services:

      a) Low Salary scales
      b) Unclear Staff Development plans
      c) Lack of Good Health Facilities , etc

iii)   Inadequate Documentation - In Program development & Maintenance.

It is realized that Inadequate Documentation in system development and maintenance is often a cause of Security failure. Without adequate documentation in your system development and maintenance, whenever a problem occurs in the system, there will be no documentation to fall on in the absence of your system guru. Even if your System guru is around, what it means is you all will make mistake whenever the guru himself also make mistake. An organization system is too big for one System guru to keep everything in his/her head.

A system is likely to grow with time, so it is always appropriate to have a good documentation in development & maintenance as and when appropriate. You can only improve your system very well if you have a good documentation.

iv)   Unauthorized Program Changes.

An organization's Computer System is so important to be allowed by an unauthorized person to effect any program changes. Organization's Program changes need to be effected by the IT Manager. In the absence of the IT Manager, his/her assistant can take up that role of program changes. It will be a security lapse for an unauthorized person to enter the Computer room to effect a program changes. In fact to effect a program change, one need the root password and this should not be known to any other person than the IT Manager and his assistant.

v)   No record of program / master file amendments

It is always advisable to keep good records of program / master file amendments. Whenever a problem occurs on the system you are likely to trace the cause of the error. Anything can happen in the running of your computer system.

vi)   Inadequate Program Backup

As good Systems Administrator, your role will be incomplete if you fail to maintain adequate Program backup. It is safer to keep good backup than rushing and not doing it well to regret later. You cannot predict when a problem will occur.

Maintaining a good backup system involves taking a backup of your system at the close of everyday. At the end of the third day, you overwrite the first one. It is appropriate to maintain a three day cycle, thus Grandfather, Father and Son cycle.

You can never belittle maintaining good backup system. I know of a story where an IT Manager of certain organization in Ghana, Africa was fired for not maintaining an adequate backup system in his organization.

vii)   Bad Physical Security:

Physical Security involves controlling who can go into areas where confidential information is kept. Computer rooms and other critical areas in the organization needs to be marked appropriately so that everyone in the organization can see that those areas are meant only to be visited by only authorized staff. In fact if those areas are not marked, unauthorized persons can visit those areas and easily can cause havoc.

viii)   Inadequate User Controls:

Inadequate User Control is a cause of Security failures. It is appropriate you maintain adequate User Controls, otherwise there is bound to be chaos in the system environment.

Users duties needs to be segregated, so that you do not have two or more staff colluding to steal the organization. Many a times in Africa, you will realize Computer room staff collude to steal the organization thousands of dollars. To avoid theft cases, it is always advisable to maintain adequate User Controls in the organization.

Chapter 4
Developing of Corporate Information Security Policy:

Every organization embarking on Information Security must have an Information Security policy. Since policies are used to define the security principles, rules and standards to which everyone must conform, the Policy document must be given its desired respect and every staff of the organization must have a copy.

The Information Security Policy is intended to help users and providers of Information Technology Services to understand what they need to know and do to make sure the company systems stay secure.

The objective of the policy document is as follows:

- To provide management direction and support for Information Security in all business units and related branches of the organization.
- To achieve common objectives and a common direction for Information Security management throughout the organization.
- Lastly to demonstrate management support and commitment for Information Security.

This important security document which must be published and promoted so that it is well- known and respected in the organization must contain the following:

- Legal and contractual compliance.
- Security education, awareness and training.
- Virus protection, prevention and detection.
- Business Continuity Planning

The policy document must be able to define specific management roles and responsibilities including making reference to an appointed Information Security Manager, who must report to the Chief Executive Officer of the organization. This senior manager must have a responsibility of Information Security in the organization.

In this Policy document, must have all responsibilities and accountability of all personnel defined. This should make reference to potential sanctions under employment contracts or prevailing computer misuse legislation where appropriate.

This policy owner, who is normally the appointed Information Security Manager, must ensure that there is annual review process defined for the security policy which involves confirming the fitness-for-purpose and making relevance of the local Information security policy. This process will have to be formally recorded.

This local security policy must be supported by lower level standards and procedures. This set of Security Standards should form the basis of the detailed documentation.

The Security policy must contain the processes for reporting security incidents in your organization. The Local Information Security Manager, must on a weekly basis send security report to the Chief Executive Officer of the Organization. In case in your organization, there is a higher Information Security Officer than the Local Information Security Manager, say a Regional Information Manager, there has to be arrangement that on a monthly basis, a report is sent to that person too.

The Local Policy must define:

- The overall objectives and scope for Information Security for that particular organization.
- The benefits to be gained in being able to share information with other branches of the Organization, external agencies, including customers, suppliers, business partners and authorities.

The endorsement of Policy Statement by the senior management of the organization must be explicit in the policy document. The policy document should be signed by the Chief Executive officer of the organization.

The Role of the Information Asset Security (IAS) Policy:

The following are some questions which should be answered to ensure that local instructions and practices are aligned to business needs and remain relevant.

- Have we been able to identify issues relating to Information Security instructions to our business?

- Will these security instructions meet our business needs, now and in the future?

- How will those instructions be maintained and kept current?

- Are these instructions adequately defined so that they are understood by our customers and Service Providers?

- Can those instructions be translated into specific and measurable service requirements?

Principles of Information Security:

What is involved in Principles of Information Security? They are normally general ideas about protecting your Organization's Information.

The following are some identified points under Principle of Information Security:

- a) There must be laid down procedures which must be implemented and maintained to protect the confidentiality, integrity and availability of your organization information.

- b) It must be clearly understood in the organization that, it is not a one man show but everyone in the organization must be involved in the Information Security deployment.

- c) Availability to organization's information must be controlled on the basis of minimum exposure needed to perform business functions while retaining sufficient open communications for effective business performance.

- d) Your organization's processes and systems for the management of Information Security must be as unobtrusive as possible and also not necessarily interfere with the conduct of the business.

- e) It is every manager's responsibility to ensure that staff and contractors of your organization know what is expected of them and that they act in a secure way to protect the company's information base.

f) A copy of your organization's policy statement must be made available to all employees and contractors. Copies should be provided on demand for consultation. New employee induction programmes should introduce and explain the policy document and related standards, procedures and instructions. Instruction guides and rules should be provided for your employees and contractors.

1) A Sample of an Information Security Policy of ABC & Co. Ltd.

Relevant Pages:

A cover page of the Policy document should have something depicting Information and in addition the organization's logo.

The First page of the Security Policy could have the Chief Executive Officer photograph signifying that the organization's management gives full support for Information Security.

Introduction: This Information Security policy, which complies with the ABC & Co. Ltd. Security Policy, is intended to help users and providers of Information Technology services to understand what they need to know and do to make sure ABC & Co. System stay secure.

Information Security:

Information Security does not concern about the security of computers and the information retained in them only, but it also extends to our documents used in the offices. ABC & Co. Ltd. relies fully on its computer and other records which are very vital for its continuation of our business. We must ensure we get ourselves secured from:

THEFT of computer equipment and that of computer and documented information which could be useful to our competitors.

DAMAGE to our computer equipment, caused either by accident or intentional.

DISRUPTION to the services of the computer as this will cause temporary loss of access to the information we need to run the business.

LOSS of the documented information that we need. Documents which you can readily locate are as good as lost.

Objectives of the Policy:

This policy:

- tells you what information security is all about and its importance to the business.
- provides and demonstrate management direction and support for Information Security in ABC & Co. Ltd. and its subsidiaries.
- tells you who the policy is for and what you need to do.
- What you need to look out for.

This policy will be supported by Information Security standards and Procedures.

Failure to comply with the policy will not be treated lightly.

This particular policy is written for all employees of ABC & Co. Ltd.

The policy covers all business information whether on paper, other storage media, processed by computer systems or data network.

It goes further to include ABC PCs, Laptop, mobile phones, scanners, E-mail and others.

Main Qualities of Information:

Information Security deals with the three main qualities of information.

Confidentiality: the assurance that ABC information is not disclosed to unauthorized person within or outside the organization and that the authorized release of information will not cause any loss or business disadvantage.

Integrity: The necessary assurance that information accurately represents the authorized business activities of ABC and is not corrupted or modified by unauthorized persons within or outside the organization.

Availability: Having firm assurance that information and information systems will be available as and when required for ABC's business needs or for compliance with regulatory or legal disclosure or any other requirement.

Ground Rules / Responsibilities:

1. It has to be ensured that clear Procedures will be implemented and enforced to protect the confidentially, integrity and availability of ABC business information.

2. It will be the collective efforts of every employee to ensure that all company information, particular the information they are responsible for, is secure.

3. Measures to protect information will be based on good risk management practice. This implies that security measures will be strictest over information or equipment whose loss would be most damaging to the business.

4. Managers will have to see to it that business continuity plans and procedures are in place and continually tested. Such strategies are designed to ensure that the business is able to operate even where there has been a security failure to a critical computer or other systems, or a major disaster.

5. Open communication should be seen as very critical for success in the business.

6. Everyone in the organization needs to be aware of information security, and must read the information security policy and carry it out. Users may only use the systems and information that they have been authorized to use to do their job.

Security Education, Awareness and Training:

All computer users in the organization must undergo information security training on their role and responsibility under the policy.

Password:

The responsibility of passwords rest with the users. Password need to be kept SECRET and managed in a secure way:

- Passwords need to be changed regularly
- You do not need to write down passwords.
- You do not need also to tell anyone about your password
- Do not use password that are easy to guess e.g. Fred, Andrea, Tim

Virus Protection, Prevention and Detection:

- Never use any unauthorized software on your organization computer system. Again never use a software from a source which is unreliable, such as the internet, free disk or games.
- Before any disk is used on the computer system, you need to scan the disk for virus.
- In case you also receive any warning of a virus presence on powering your computer, shut down your system immediately and contact the Help Desk.

Hardcopy:

All Information on paper need to be kept safe. Printout of reports must be collected off the printer as soon as possible. The reports need to be stored in drawers and cabinets. Confidential and/or sensitive information should be locked away. If any report is not needed, it needs to be shredded.

Unattended Terminals:

When you are logged into the computer, it is your duty to ensure that no one else uses the systems you are using. You don't need to leave your computer unattended when it is logged on or unprotected.

Copyright:

ABC Company will only use authorized software. Software covered by copyright must never be copied without the owners consent. This is considered illegal and will result in disciplinary action,

Back Ups:

Back up copies of all information held on personal computers and carried out on a regular basis. Label backups and store them in a safe place. It is always appropriate you keep backup disks and tape away from heat, smoke, food or drink.

Physical Security

Adequate physical security measures must be taken to protect computer systems and information from theft, damage and misuse.

- Where possible, use blinds or other screens to stop outsiders from seeing in.
- You must always avoid eating, drinking or smoking near equipment
- Lock equipment away, when not in use.
- Make sure all portable PCs, mobile phones and other moveable equipment are protected from loss or theft. Do not leave such equipment in hotel rooms, in cars or on luggage racks.

Using External Systems.

The organization's computer services must not be used to access information which is:

- of a discriminatory or harassing nature
- derogatory to any individual or group.
- Obscene or X-rated
- Of a defamatory or threatening nature

It is the responsibility of the Information Security Manager acting in conjunction with the IT Department, to ensure that all Computer users in the organization comply with the standards in the Electronic and Media Services Policy.

Confidential Undertakings

All employees must sign a confidentiality undertaking, agreeing to safeguard the organization's information and information assets against prejudicial interests or other interests while within employment or any time afterwards.

A Security Breach occurs when Information Security measures have failed to work as required for designated situations.

A Security Incident is any circumstance or activity capable of causing or which might have caused, the breakdown of the normal course of information flow within ABC & Co. in contravention of Information Security rules, regulations and responsibilities.

All employees of the organization are to ensure that Security Breaches and Incidents are reported quickly to the Information Security Manager or Help Desk and the head of Department.

Examples of Security Incidents and Breaches

- Unauthorized removal or "borrowing" of information from the company premises.
- Computer Viruses
- Information Stolen and given to a supplier or competitor.
- Missing files
- Theft or loss of equipment
- Disappearance of sensitive data
- Login not working
- If the data/time of last login is incorrect.

ABC & Co. management is committed to ensuring that the quality of Information is outstanding.

In order to ensure that these basic responsibilities are made clearer, the Local Information Security Office will produce standards and procedures and other necessary documentation which would help the process and "sell" the organization's business.

POLICY REVIEW

In case you have any comments on the Policy, please send them to the Information Security Officer or Manager.

The Policy will be updated on an annual basis. Also annually, a review of the policy document will have to be carried out.

2) A sample copy of Electronic Media and Services Policy of ABC & Co. Ltd.

CONTENTS

Appendix A:

Appendix A
Version History

Version / Status	Release Date	Comments
1.0/Approved	02/02/2006	Approved by CEO of ABC Co. Ltd.

1.0   Introduction

1.1   As an advanced technology company, we increasingly use and exploit electronic forms of communication and information exchange. Employees have access to one or more forms of electronic media and services (computers, e-mail, wire services, telephones, fax machines, external electronic bulletin boards, on-line services, the Internet and the World Wide Web).

1.2   The Organization encourages the use of these media and associated services because information technology is our business, because they make communication more efficient and effective, and because they are valuable sources of information, e.g. about ABC's customers, vendors, new products and services. However, the Organization emphasizes that these media and services, like all other forms of Organization property, are for Organization business and not for personal use.

1.3   This policy outlines the general principles to be applied by ABC personnel and contractors to ensure the proper use of electronic media.

1.1 **Policy**

1.2 Electronic media must not be used for transmitting, retrieving or the storage of communications of a discriminatory or harassing or X-rated, or which are derogatory to any individual or group, or for "chain letters" or for any other purpose which is illegal or against Company policy or contrary to the ABC organization's interest.

1.3 Electronic media must not be used for personal profit or gain.

1.4 Electronic information created and/or communicated by an employee or contractor working on ABC Co. premises using e-mail, word processing, utility programs, spreadsheets, voicemail, telephones, Internet/BBC access, etc., will not generally be monitored by the Company. However, the Company reserves the right, at its discretion, to review the electronic files and messages of any individuals working for ABC or on ABC premises to the extent necessary to ensure that electronic media and services are being used in compliance with this law and with this and other ABC policies.

1.5 Employees and contractors must respect the confidentiality of other persons' electronic communications and must not attempt to read, "hack" into other systems or another person's login, or "crack" passwords, or breach computer or network security measures, or monitor electronic files or communications.

1.6 No "on-line" service (e.g. dial-up Internet access via Africanonline or any service provider), accessed via dial-up or any other facilities, may be connected to any ABC equipment without the express permission of the IT Manager and Chief Executive Officer.

2.6 No e-mail or other electronic communications may be sent which attempt to represent the sender as someone else or another company.

2.7 No directly connected "on-line" service may be connected to any ABC equipment without the express permission of the site IT Manager and Chief Executive Officer .

2.8   No Branch in ABC Co. or its headquarters may implement a WWW presence on the Internet without the express written permission of the site <u>IT Manager</u> and <u>Chief Executive Officer</u>.

2.9   Electronic media and services must not be used in a manner that causes network congestion or significantly hampers the ability of other people to access and use the system, e.g. by sending large files across ABC net, thereby potentially delaying business critical data (such as IT information).

2.10   Anyone obtaining electronic access to materials (such as information) belonging to other companies or individuals must respect all copyrights attached to such materials, and must not copy, retrieve, modify or forward materials protected by copyright except as permitted by the copyright owner. Public domain software may be retrieved from the Internet only with the express permission of /the IT Manager and only when the appropriate virus protection procedures have been followed.

2.11   Any messages or information sent by an employee to one or more individuals via an electronic network (e.g. bulletin board, on-line service, e-mail or Internet) are statements identifiable with and attributable to ABC. Network services and World Wide Web sites can, and do, monitor usage and can identify which company, and often which individual, is accessing their services. It must be reiterated, therefore, that all communications sent by employees via a network must comply with this and other Company policies, and must not disclose confidential or proprietary ABC information.

2.12   Any employee found to be abusing the privilege of Company-facilitated access to electronic media may be subject to disciplinary action, depending on the seriousness of the breach of policy. In extreme cases, such action may result in termination of employment or contract.

2.13   With the high incidence of viruses associated with transmitted data in the PC environment, users must ensure that the latest version of an anti-virus software (e.g. McAfee or Kaspersky) is installed on their PCs. From time to time users must scan their hard disks and other storage media like tapes, pen drives, CDs and others.

CHAPTER FIVE
DATA SECURITY AUDIT:

Data Security Audit is very critical in Information Security. That is the only way we can determine whether, we are actually putting into practice what is expected of us by way of conforming to the rules and standards in Information Security. Data Security Audit, will enable us to assess our own controlled environment. Through the Data Security Audit we will also be able to measure the overall progress made in improving the level of security in our business units within the organization.

Since we can never have a perfect environment, it is only through Data Security Audit that we will be able to know our deficiencies and try to improve upon it. Data Security Audit involves using prepared questions by the Information Security Manager to carry out a check on the control environment, whether the Information Security standards and procedures have been implemented.

Data Security Audit comes into 2 main forms depending on the size of your organization. You can decide to have the normal Audit programs of your organization carried out by your Internal Audit Department and supervised by the Information Security Manager, or the second aspect which is by using Self-Assessment Process. This Self - assessment process which can be carried out, say on a quarterly basis will be carried out throughout the organizations subsidiaries or if it is a multi-national company, throughout the various countries which form the organization. In all cases it is advisable to prepare questions covering all aspects of Information Security like the following for the audit:

- Security Policy
- Security Organization
- Asset Classification and Control
- Personnel Security
- Physical and Environmental Security

- Computer and Network management
- System Access Control
- System Development & Maintenance
- Business Continuity Planning
- Compliance

Self – Assessment Process:

The following is a sample questionnaire of Information Security Self – Assessment:

Security Policy:

1. Has a Local Information Security Policy been developed and implemented ?

....

Security Organisation

2. Has your organization appointed Information Security Manager? Name: ....

3. Do all Computer Applications Systems have an owner identified?

....

4. Are Information Owners, Users and Custodians made aware of their responsibilities through a security education and training process?

....

5. During the annual budget process, is provision made for the implementation of Information Security requirements?

....

Assets Classification and Control

6. Is your organization data classified according to the Confidentiality guidelines?

....

7. Does your organization maintain an inventory of Information Assets?

....

8. In your organization, has there been a documentation of individual assets owners?

....

Personnel Security

9. In your organization, do you have a system where employees and contractors sign a confidentiality agreement prior to their initial connection to ABC's IT facilities?

....

10. In your organization do you have an ongoing Information Security education and awareness programme for your staff ?

....

11. Is there a local procedure for reporting Information Security incidents to your site Information Security Manager?

....

Physical and Environmental Security

Do you have the following areas secured either in a separate building or located in the main building in a strictly controlled area with access to staff with pre-authorized business need? Are these particular areas locked and access to them monitored?

12. Organization's Telecommunications room?

....

13. Cabling connecting computer equipment and networks?

....

14. Areas where sensitive files are kept e.g. back-ups, tape libraries, disk Storage areas?

....

15. Do you have appropriate fire extinguishers located in computer room areas?

....

16. Are smoke detectors installed in the computer room?

....

17. Do you have a "clear desk" policy in operation to ensure that sensitive information is not made available to any person with access to premises out of normal hours?

....

Computer and Network Management

18. In your organization, do you have a virus protection scheme and procedures in Place to prevent or minimize the risk of a virus being installed or becoming active?

....

19. In your organization, do you have all your PCs equipped with the standard approved virus protection software?

....

20. Do you have a system where the virus – protection software is kept up-to-date?

....

21. Do you have effective procedures and responsibilities established for reporting virus incidents?

....

22. Do you have effective documented operating procedures for the following:
    a. Start – up and close – down ?   ....
    b. Data backup?   ....
    c. Testing and verification of backup media? ....

23. In your organization, has there been an effective segregation of duties between operations, system development, and end-users?

System Access Control

24. In your organization, are there all privileged (super-user) facilities identified and documented?

....

25. Is access to specific applications or systems facilities effectively controlled so that only properly authorized individuals have access to the appropriate information necessary to perform their job function?

....

26. In your environment, are privileged facilities allocated to individual on a need –to-use basis?

....

27. In the disposal of discarded magnetic media or printed copy, are appropriate measures taken to ensure it is done in a secure manner?

....

System Access control

28. In the access to specific applications or system facilities, are they effectively controlled so that we have only properly authorized individuals having access to the information necessary to their job function?

....

29. Are all privileged (super-user) facilities identified and documented?   _____

....

30. Are the allocations of privileged facilities subject to an authorization process and operated under peer-supervisory controls?

....

31. Are identification and password unique to each user?

....

32. Do you normally have passwords set to non-display?

....

33. Is there a recommended length and content standard for passwords?

....

34. Are required security controls included in the systems documentation?

....

Are there formal change control procedures which include the requirements that:

35. Changes are accepted by the authorized User before full implementation?

....

36. Do you have an audit log maintained for the history of all change requests?

....

Business Continuity Planning

In your Organization do have a Business Continuity Plan written?

....

37. Do you have Business Plan tested on a regular basis?

....

38. Is there a technical recovery plan which covers the restoration of the IT facilities, including the provision of alternative hardware?

....

Compliance

39. Are there regular audits of software in use in order to identify any unauthorized or non-standard software?

....

40. Are there guidelines on the retention, storage, handling and disposal of organization's records and information which meet statutory require-ments and support the business activities?

....

41. In your organization, does the handling of data concerning individuals person comply with the local data protection or privacy laws?

....

Information Security Self-Assessment – Completion Instruction & Control Exemption Form.

In trying to complete the Self – Assessment Questionnaire, it is very appropriate you go through the Completion Instruction. This completion instruction is normally attached to the main questionnaire to explain most of the responses which you will have to provide for the main Self-Assessment questionnaire.

Regarding the answering of the Self-Assessment questionnaire, it is very appropriate that both the Information Security Manager and the IT Manager do it together. In bigger organizations or multi-nationals where you will have a number of local Information Security Managers and a regional Information Security Managers coordinating the activities of various Local Information Security Manager, all completed Self-Assessment questionnaires are finally submitted to Regional Information Security Manager.

Each question in the self- assessment document will have to be answered with the following responses:

"Y" – Process or activity in place. Please also try and indicate:

   "1" - Totally effective

   "2" – Requires minor improvements.

   "3" - Require considerable improvement

"N" – The control is not required or not applicable to the local environment. The risk has been assessed and exemption has been approved by the Senior Manager of the business units.

"P" – In plan.

The above codes should be entered in the left - dotted area.

In case there are any questions assessed as "Y-2", "Y-3" or "P", there is always the need to enter a date in the right - dotted area to indicate the planned date for implementing a totally effective control.

For each of the various questions of the Self-assessment questionnaire that are answered with an "N", a corresponding Control Exemption Form must be completed to indicate that the risk of not applying the control is acceptable.

Signatories:

Both the completed Self-Assessment Questionnaire and Control Exemption Form will have to be signed by Info. Security Manager, I.T. Manager and General Manager of the local site. This is the only way that, it can be certified that the questionnaire has been completed by observing all the required rules and regulation.

Information Security Self-Assessment – Control Exemption Form:

A control self-assessment rating of "N" indicates that the recommended control has been assessed and found not to be required or not applicable to the local environment. The Senior Manager of the Business Unit should approve the claim for exemption from the control.

The following form should be used to justify the assignment of an "N" rating.

Self – Assessment question number: ....

Reason for "N" rating:

... Non viable with technology in use	Describe technology:
... Risk reduced by other means	Indicate compensating controls
... No perceived threat	Justify:
... Other reason:	

Approval	Name	Signature	Date
Info. Sec. Mgr.	....	....	....
IT Manager	....	....	....
Chief Exec. Off	....	....	....

Compliance with legal requirements:

Your study of Data Security Audit will be incomplete without mentioning compliance with Legal requirements. Under Compliance with Legal requirements, we have the following:

a) Control of Proprietary software copying.
b) Safeguarding of company records
c) Compliance with data protection regulation.

Control of proprietary software copying:

Control of proprietary software is so important that it is very critical we take a special look at the standards involved. The following are some of the standards we need to observe:

1) PCs users must be advised that it is against your organization's policy to make unauthorized copies of proprietary software.

2) Your organization's personnel must be instructed not to use software packages they themselves have not obtained on company business.

3) There must be regular (at least bi-annual) audits of software in use in order to check the software register and identify unauthorized software.

4) Computer users must be clearly instructed not to copy software, or to use copied software, outside the terms of the licenses (and advised that such actions may be criminal offences).

5) Statements of policy must be made on the use of copyright-protected software (stating compliance with any prevailing national legislation regarding copyright).

6) In your corporate organization, it will be advisable to have software registers of all authorized copies of software.

7) Those responsible for software procurement and licensing should be known to users so that extensions to licenses or obtaining permission to make additional use proprietary software can easily be managed internally.

Safeguarding of company records

Since we cannot compromise on our quality of records, it will be advisable to ensure that vital records of the business unit are protected from loss, destruction or falsification.

Compliance with Data Protection regulation.

The handling of data is also important that, it will be very advisable to comply with prevailing Data Protection or Privacy laws.

Chapter Six
User Access Control

User Access Control plays very useful roles in our day to day management of our organizational computer system. To have a healthy corporate system running, it is very prudent that we put appropriate strategies in place.

What policies, standards and procedures do we put in place so that there is orderliness in the running of our system? By the close of the chapter, the following are the areas which will be covered.

- User access management
- User responsibilities
- Business requirements for system access
- Computer access control
- Application access control
- Monitoring system access and use.
- Network access control

6.1 User access management:

User access management involves putting adequate management controls in place so that users can enjoy the use of the system. Issues like which forms users need to fill before they are granted access to the system. Review of user access rights are some of the issues which will be discussed under this session.

The following are some of the areas which will be discussed under this session:

- User registration
- Review user access rights
- Unattended user equipment

User Registration:

In the proper management of a corporate system, it has to be established appropriately what information, a user can access.

The following are some of the standards which must be observed under User's Registration.

1) Users of the system must have documented authorization from the Information owner for the use of the service.

2) The various users of the system must be given a written statement of their access rights.

3) Also users must be required to sign an undertaking to indicate that they understand and accept the condition of access.

4) There must be appropriate procedures to ensure that redundant user IDs are not re- issued to other users.

5) There must be a formal record maintained by the Information Custodian of all persons registered to use the service.

6) Also accounts must be removed or changed immediately for users who leave the company or change jobs internally.

7) Proper procedures will have to be established for the notification of the Information Custodian of those who leave the organization.

Review of User access rights

From time to time, it is very appropriate for Information Owners to check for who has access to their business information. Those users who mistakenly use certain business information will have to be corrected.

The following are some of the standards which will have to be established:

- There must be a formal process to review users' access capabilities at regular intervals.

- A formal process will have to be established for checking privilege allocations at regular intervals to ensure that unauthorized privileges have not been obtained.

Unattended user equipment:

Under this we normally advise users on good practice in shutting down terminals, workstations or personal computers at the end of working session or when temporarily left unattended.

The following are some of the standards or procedures which must be observed.

- Instructions will have to be given to users to terminate their sessions when they have finished working.

- Also users must be instructed to secure their PCs with lock after close of work.

- Users must again be instructed to log – off from the remote computers or host PCs at the end of each session.

6.2 Users Responsibilities

This area is a very critical since issues like using Passwords effectively will be handled. Since Password is comparable to a key for a door, it needs to be guarded well, otherwise you will have thieves breaking into your system.

The following are some standards which will have to be observed:

1) Users must be instructed to keep passwords confidential.

2) Also users must be instructed to ensure that individual passwords are selected to maintain accountability.

3) Whenever any users experience any indication of possible system or password compromise, that user will have to effect a change immediately.

4) Also users must be instructed to select passwords with a minimum length.

   (a minimum of six characters is recommended).

5) Users must also be advised to avoid keeping paper records of passwords.

6) For safety, it is advisable to change passwords at regular intervals (one month recommended).

7) Normally users must be instructed to avoid using password connected with or based on:

   a) Family names, initials or car registration numbers.

   b) User IDs, usernames, group IDs or other system identifiers

   c) Telephone numbers (or similar all-numeric groups)

   d) All numeric or all alphabetic characters.

   e) Month of the year, days of the week, or any other aspects of the date.

8) Users must be instructed to change temporarily passwords as soon as they log on with the password.

6.3 Business Requirements for System Access

In fact this aspect is concerned with drawing up a policy for access to business Information. Since it is appropriate to control access to Business Information, standards and procedures will have to be put together.

Under Business Requirements for System Access, we tend to consider, Documented access Control Policy, Privilege management and User Password management.

Documented Access Control Policy:

The following are some aspects of Standards and Procedures under documented Access Control Policy, which needs to be observed.

1. Information Owners will have to provide clearly defined and documented access policy statements, giving definitions for access rights and restrictions for users and groups of users.

2. Regarding Information dissemination and Entitlement, "Need to-know" rules will have to be applied.

3. All Business Applications must have documented statements of access control policy.

4. There should be definition for standard user access profiles and used for common categories of job.

5. Access control policies must deal with contractual, regulatory and legal requirements to protect access to data or services.

Privilege Management

Under this we normally consider, how some employees enjoy greater and easier access to the business information.

The following are some Standards and Procedures which needs to be observed under this section.

1. All associated privileges with each system feature (e.g. operating system, database management system) and different categories of staff which needs to be allocated must be identified.

2. It is always advisable that, you do not allocate any privileges until the authorization process is complete.

3. A record need to be kept for all privileged user activity.

4. The use of privileged access authority must be supported by peer supervisory controls and supervisory review of logs.

5. Privileges will have to be allocated to individuals on a need to use also event by event basis.

User Password Management:

_ Under User Password Management of Business Requirements for System Access, we normally consider issues like ensuring people use passwords effectively.

The following are standards and procedures considered under User Password Management.

1) All users must sign an undertaking to keep personal passwords confidential and work group passwords (where these are necessary) solely within the members of the group.

2) Users must always acknowledge receipts of passwords.

3) Conveyance of passwords through third parties, or unprotected electronic mail messages will have to be avoided.

4) It must be provided with a once only temporarily password when being initiated to the system or recovering from a lost password.

6.4 Network Access Control:

Under Network Access Control, the following are the area which needs to be convered:

- Limited Services
- Enforced path
- User authentication
- Segregation in Network
- Remote Diagnostic Port Protection.

Limited Services: This involves formalizing the authorization of users to connect to corporate or public network services.

The following are some of the procedures and standards under Limited Services:

- Users must be provided with access to the services that they have been authorized to use.
- Some users may have to be denied access to sensitive applications in certain circumstances, especially remote access from unknown locations.

Enforced path:

Under Enforced path, we normally consider the controlling of the route between the user terminal and accessed service.

The following are some Standards and Procedures which needs to be observed under Enforced path.

- Controls will have to be established to restrict the route between the user terminal and the computer services.
- If possible, dedicated lines and telephones numbers should be used whenever viable.
- There should be a practice of automatically connecting ports to specified applications systems or security gateways.
- Menus and sub-menus will have to be customized for individual users.
- Measures will have to be adopted to restrict or prevent network roaming and browsing.

User authentication:

Checks will have to be established for the identity of users accessing ABC Information through uncontrolled networks.

The following are some Standards and Procedures under User authentication.

1) The business risks of users connecting by public or third party network must be evaluated.

2) The identity and authority of users connecting by public or third party networks must be authenticated at application, computer or network level.

   The following authentication techniques may be applied:

   a. Challenge/ response

   b. Line encryption

   c. Private line or network user addresses (NUA) checking.

3) All access attempts must be logged and reviewed regularly against authorized access lists.

Segregation in Networks

The essence of Segregation of Network is to create network domains. Gateways and firewalls will have to be provided for secure paths between the domains.

The following are some standards and Procedures which needs to be considered:

- Large networks should be divided into smaller domains to help keep security management, administration and access control manageable.

- The use of domains, firewalls and gateways will have to be defined and supported by information security procedures.

- Also all domains must be covered by security procedures and security management.

6.5 Computer Access Control

Under this we normally concern ourselves with issues like terminal identification, Terminal log-on procedures, User identifiers, Password management system and terminal time-out.

Terminal Identification:

Under this section, we treat issues like checking that computers run from their correct terminals.

The following are some of the standards and procedures we treat under this section.

- Highly sensitive application will have to be protected by automatic terminal identification and authentication.
- Terminals or modems which are subject to automatic identification must have adequate physical protection.

Password Management System:

Under this section we concern ourselves with the proper management of the use of passwords. We further ensure that provision of effective, interactive facility which ensures quality passwords is provided.

The following are some Standards and Procedures which will be considered under this section:

1) All users of computers must have their own unique ID and password.

2) "Guest" passwords will not be entertained.

3) Also regarding the changing of passwords, it must be forced at regular intervals, say on a monthly basis.

4) The password management system should detect and prevent the re-use of recent passwords or significant parts of recent passwords.

5) Access failure or refusal must be recorded as security events.

6) The system of Password Management must discourage the choice of very weak passwords (easily guessed or detected by trial or error.)

7) The password management system must ensure that passwords being changed are not displayed on screen, but that confirmation (probably by re-keying) is performed.

8) There must be Quality checks to detect and prevent the use of recognizable string of characters in passwords e.g. days or months, company names, user IDs, user names, etc.

6.6 Application access control:

Under the Application access control, we normally handle issues like setting up controls within the application to limit access.

The following are some standards and procedures which can be considered under this section.

1) An access policy for each application will have to be established. This policy will have authorized by the application owners and it will have to be in line with the Information Security policy.

2) On a regular basis, the applications' access policy will have to be reviewed by Information owners and Independent authorities like Internal Audit.

3) There should be menus to be used to control access to the application system functions in accordance with the application access policy.

4) It has to be established that users are denied knowledge of data or systems function which they are not authorized to access.

5) There must be controls to ensure that outputs from applications which handle sensitive data contain only the data that is relevant to the use of the output.

6.7 Monitoring System Access and Use.

This area of System Access control normally concerns with the keeping records of unusual events, specifying what kind of monitoring should be done and setting accurate time on the computer.

Under this section we normally concern with the following:

- Event Logging
- Monitoring System Use
- Clock Synchronization.

Event Logging:

The following are some standards and procedures under Event Logging.

1) It must be ensured that Audit trails which record other security relevant events must be taken and retained.

2) It must be arranged that audit trails are kept for an agreed period (accepted by the organization's legal office).

3) Audit trails which normally records exceptions must be taken and retained.

4) Normally audit trails of all third party access must be taken, reviewed and retained.

Monitoring System Use:

Under this we normally consider, specifying what kind of monitoring which need to be done.

The following are certain standards and procedures which must be observed under this.

1) In your organization, there must be formal authorization procedures for monitoring systems use.

2) Normally, routine monitoring must include checking access failures.

3) Requirements for monitoring system use must be determined by risk assessment.

4) There must be routine monitoring which will include tracking selected transaction.

Clock Synchronisation.

_This section normally involves setting accurate time on the computers.

The following are some standards and procedures which must be considered under this section.

1) In your organization, there must be an established responsibility for Setting and monitoring systems clocks.

2) You must have procedures for setting and re-setting clocks to date and time (BST/GMT or equivalent). This must be clearly documented and established.

3) Access to clock adjustment function must be limited to a minimum number of authorized staff.

4) Logs and audit trails must be checked to ensure that they are correctly time stamped at the time of recording.

www.ingramcontent.com/pod-product-compliance
Lightning Source LLC
Chambersburg PA
CBHW051216050326
40689CB00008B/1335